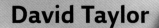

Collins

The Secret Life of Cats

David Taylor

First published in 2006 by Collins
an imprint of
HarperCollins Publishers
77–85 Fulham Palace Road
London W6 8JB

www.collins.co.uk

10 09 08 07
6 5 4 3 2 1

A catalogue record for this book is available from the British Library.

David Taylor asserts the moral right to be identified as
the author of this work.

Editor: Heather Thomas
Designer: Rolando Ugolini
Photographer: Mark Read

ISBN-13: 978-0-00-724475-1
ISBN-10: 0-00-724475-4

Colour reproduction by Dot Gradations Ltd, UK
Printed and bound in Malaysia by Imago

Can cats really
read your mind?

Are cats mind readers or do they understand when you say, 'I've got to take Grimalkin to the vet again this morning'? Some vets and animal psychologists believe so. Many vets report that people often cancel their appointments because they cannot find the patient. I know one vet who has stopped taking cat appointments altogether because there have been so many 'no shows' in the past!

Home...
sweet home

If you want a cat as a pet, you can't beat going to a cats' home and picking the first needy, cross-bred moggie who looks you in the eye. However, if you really desire a pedigree, which breeds have the best temperaments for family life? The best ones are Persians, Birmans and Shorthairs of various kinds. Not quite so good are the Siamese, Maine Coon and Norwegian Forest Cat. Least family-fond is the Korat.

Good, good, good...
good vibrations

Because cats are acutely sensitive to vibrations, they will regularly give warning of impending earthquakes or any volcanic activity. Strange behaviour by house cats was widely reported for up to fifteen minutes before the Agadir, Skopje and Alaska disasters in the 1960s, and peasants living on the slopes of Mount Etna in Sicily know it is best to follow when a drowsy fireside tom suddenly scarpers for no apparent reason!

How do cats
express themselves?

The facial markings of cats are quite useful in emphasizing the expressions their mouth, eyes and ears form, particularly threatening ones. Body markings also help to accentuate signals sent out by a cat's body posture. Whether slight differences in pattern enable other cats to visually identify an individual is unknown, but it is a distinct possibility.

Does your cat
touch you up?

Touch is another means of cat communication. Cats do it by rubbing their noses on and pressing their bodies against each other, and by grooming other cats. Delightfully, they do it to us humans, too. My Birmans specialize in making friendly bumps with the head. These behaviours signify affection and contentment and also reinforce the bonding that is so important in their lives. Of course, they also do this touching when they are 'touching me up' for something they want!

Sending smelly
messages

Scent marking is important to cats. Leaving their own personal dollops of pong is a way of sending messages as well as defining territorial boundaries, ownership of property and the status of the cat doing the marking. Secretions from glands on the cat's chin, urine and faeces all carry these scent messages. Cats depositing faeces indoors, on a bed, have often been stressed in some way and are trying to reinforce their bond with their human family.

Tigers in
miniature

The hunting technique of cats is as perfect
as that of the tiger. Having spotted his prey,
Puss immediately takes advantage of any
available cover. Pressing his body close to the
ground, he darts forwards, before a pause in
ambush, still flattened, and then another 'slink
run'. Another pause, treading movements with
hind feet, tail twitching, eyes at their widest.
Now he breaks cover, sprints and pounces, hind
feet planted firmly on the ground for stability.

Cats bearing
gifts

You have, no doubt, been on the receiving end when your cat marches proudly into the house bearing a mouse, bird or mole. He isn't merely flaunting his hunting prowess but giving you, a valued family member, a gift. It's in all cats' natures – bringing food for the young or adults of the opposite sex. Siamese toms love to do it for their kittens. So, don't scold your pet and don't imagine he's underfed.

What big teeth
you've got!

The biting mechanism of all cats, from tigers to our fireside familiars, is highly sophisticated. The distance between the fang teeth is adapted to deal with, by lethal dislocation, the neck vertebrae of their principal prey species – for Moggy, that's the mouse. The teeth possess ultra-sensitive nerve receptors that signal when they are in the correct position and then the jaw muscles contract unusually quickly. Crrrunch!

All coat colours
and patterns

The wild relatives of our domestic cats sport coat colours and patterns that are designed to function as camouflage. Some pattern features, such as the spots on the back of tigers' ears, act as signalling equipment. Our domestic cats come in a bewildering range of colours and designs, mainly due to years of selective breeding by Man, but sometimes, as with this cat, traces of the tabby patterned coat of its African wild cat ancestor can still be seen.

Celestial navigation
...do cats do it?

Cats can sometimes find their way home over surprisingly long distances. The record journey was that of a cat who travelled 1,400 miles from California to Oklahoma in the USA. It took the cat fourteen months but he was positively identified on his arrival by X-rays of an old hip deformity. So how do cats do it? Probably by celestial navigation – noting the position of the sun by day and the stars by night, in conjunction with an internal biological clock.

What's mine
I scratch!

Cats and owners don't see eye to eye when it comes to scratching, particularly indoors, and even more so when an antique armchair is involved! Cats don't scratch 'to sharpen their claws'. Although scratching helps to tidy up the old outer layers of claws and exercise the foreleg muscles, its main purpose is marking territory. What's mine I scratch! Insecurity and stress can increase a cat's bouts of scratching.

It's raining
cats... yes, really!

When I used to live in Glasgow, cats were often brought to my veterinary surgery with fractures after falling from high tenement window-ledges where they had been dozing. The fracture rate increased, the higher the window-ledge, up to seven storeys. Strange but true – above that, it began to decrease! Why? Well, once a falling cat reaches maximum speed he relaxes and spreads out his legs like a freefall parachutist. Relaxed bodies are much less likely to fracture.

Read my body
language... or else!

Trouble could be brewing here! It's all in the body language. The aggressive cat (on the right) has his ears pricked and laid back, his mouth open wide and snarling, his whiskers bristling forwards and his tail swishing from side to side. The other cat, in defensive posture, stands with his back arched and body turned at an angle, hopefully trying to make himself look bigger!

Why do our cats purrrrrr?

A cat's purring can have a number of functions: it can act as a bonding signal between mother and kittens, as a 'dinner gong' summons for suckling to begin, as part of a female's mating display when she's on heat, and an expression of well-being when being stroked and all is well with the world. More oddly, purring can be heard sometimes when cats are in extreme pain.

Cross-breeding
with wild cats

Do our domestic cats ever breed with wild cat species? Yes, occasionally. The beautiful Bengal here is a hybrid cat produced by breeders crossing a domestic shorthair with an Asian Leopard Cat. The mysterious Kellas cats, long-legged black animals living in the forests of Morayshire in Scotland, have been proved by DNA sampling to be the result of Scottish wild cats mating with domestic cats without the help of breeders.

The eyes
have it...

Cats' eyes can signal a wide range of emotions. When a cat is in an aggressive mood, his pupils are closed to slits, whereas when he is on the defensive they are wide open. Most attractive of all – at least, so Puss thinks – are the wide eyes, alert and gazing intently at the object of their desire, of a cat importuning his owner – and usually succeeding in getting what he wants!

Performing a
balancing act

The cat's tail acts as a counter-balance when teetering along fences, a kind of rudder when leaping and a counter-weight swung away from the direction in which the body is travelling when cornering at speed. No wonder cheetahs, those fastest of cats, have such long tails. But exceptions prove the rule! Cats with very short tails, or no tails at all, such as the Manx cat and lynx, seem to cope very well with jumping and balancing.

Turned on...
and tuned in

Cats frequently appear stone deaf when you are telling them something they don't want to know! In fact, their sense of hearing is superb. Thirty muscles (as compared to just six in us humans) move the ears to focus on sounds. Over 40,000 nerve fibres link a cat's ear to his brain, compared with 30,000 in us. They are particularly tuned to high-frequency sounds that are produced by high-pitched squeaks and trills. I wonder why?

Cat conclaves...
what are they?

Masonic lodges for moggies? Prayer meetings for pussies? In any neighbourhood where there are cats, some areas are not claimed by any individual animal as its territory. This 'common land' is used from time to time for meetings. The local cats will gather round peacefully and just sit silently looking at one another. So what are these assemblies really all about? Do the cats simply enjoy being together? Or do they converse telepathically? We do not know.

How bright
is your cat?

How do you measure the intelligence of a cat?
With great difficulty. Many scientists use a
method comparing the weight of the cat's brain
to the length of their spinal cord. This ratio
indicates how much grey matter controls how
much body, and should be bigger in more
intelligent species. A human gives a ratio of
50:1; a cat a ratio of 4:1. So does that mean
we're about twelve times brighter than Puss?

The flehmen
phenomenon

You will sometimes see a cat make a rather curious grimace, lifting his head, opening his mouth a little and wrinkling his upper lip and nostrils. This so-called 'flehmen' phenomenon is a sort of super-smelling process, which involves a special gland in the roof of the mouth and is often used when the cat sniffs at a patch of urine or some strongly smelling substance. One of my Birmans does it when he's sniffing whiffy Gorgonzola cheese!

Do cats have a
sweet tooth?

If you, like me, are on the staff of a fussy feline, you know only too well that cats do have a fine sense of taste. Until recently, however, it was thought that cats, unlike dogs, did not possess the ability to recognize sweet flavours. Now we know that a few 'sweetness-detecting' nerves link the cat's tongue with its brain – and the numbers of them seem to be increasing!

Can our cats
talk to us?

Cats are talkative creatures. Each cat has his
own individual voice, recognizable by friend
and foe, which he uses to communicate with
his fellows as well as human beings. Cats can
produce about three dozen sounds, including
nineteen different types of 'miaow'. Indeed,
their 'chatter' sound, often heard when they
spy a bird nearby, may even be capable of
luring their hapless prey to come closer!

The agony
and the ecstasy

Many cats get extremely turned on by the smell of catnip and valerian in the garden, rolling in the plants in obvious ecstasy. So do wild cats, such as lions, tigers and lynxes – although not in your garden normally! These plants contain a chemical, nepetalactone, which is closely related chemically to a compound secreted in the urine of a female cat when she is in heat.

Why do cats
have whiskers?

What is the point of whiskers? So say we
blokes when we are shaving on a winter's
morning. But for cats they are extremely
important, and removing them can markedly
disturb a cat for some time. No, they aren't
used as a measuring device to check whether
the animal can pass through a given space,
as many people think. They serve as highly
touch-sensitive antennae to locate and
identify things close by in darkness.

What do cats
like to drink?

I've known the odd cat who liked to lap bitter
beer occasionally, probably because of the
distinctive yeasty flavour. Many cats drink very
little, particularly if they are fed on canned food,
but, nevertheless, they must have fresh water
available at all times. It's funny, though, how
often cats who are given sparkling clean water
prefer the taste of greenish rainwater puddles
or the contents of flower vases!

Can cats see
in the dark?

Although cats cannot see in total darkness, their night vision ability is extraordinary. Their eyes can gather and magnify the faintest glimmers of light. Behind their retinas are light-intensifying screens composed of glittering crystals. These reflect light that has passed through the photo-receptor layers of the eyes and are the reason that cats' eyes gleam so dramatically. Hence the name 'cat's eyes' for the reflective discs in roads.

Are cats really
sensitive to smell?

Unlike their cousin the tiger, which is thought to have little or no sense of smell, domestic cats are talented sniffers with four times as many smell receptors in the lining of their noses as us humans, but around seven times fewer than a long-nosed dog. They are particularly sensitive to nitrogen compounds in odours, such as those emitted by food that has gone 'off'. Puss is a fussy creature and always insists on *fresh*!

A wild cat
in your living room

It is fairly certain that today's fireside moggy is descended from the African wild cat, an animal that looks very much like a long-legged tabby with a smallish head – the sort you might see any day in your neighbourhood. Generally happy living close to human settlements, the cat began its process of domestication millennia ago when it was attracted to the vermin that were to be found in early man's grain stores.

It's a cat's
life... snoozing

On average, Puss spends sixteen hours out of twenty-four asleep (up to eighteen hours when he is old) although this long snooze-time may be composed of a number of 'cat naps'. So why do cats need so much sleep? Probably because their essential nature as hunters requires them to conserve maximum energy for a few brief, high-performance bursts of activity.

The thrill of
the chase

Cats who regularly go a-mousing are not trying to tell you that they could do with more food. The best mousers are those with full stomachs, as farmers who keep cats in their barns know well. No, it's the pure thrill of the chase, the built-in urge to hunt, inherited from their wild ancestors, that motivates them. For the modern Puss it's much more fun to hunt a mouse or a bird than a dish of Whiskas or Felix!

Fruit and
veg are out!

Although you may well know a cat (I bet it's a
Siamese!) who is partial to the occasional grape
or slice of tangerine, cats, unlike us humans
and many other species, do not need Vitamin C
in their diet to stay healthy. Having evolved
as a pure meat-eater, the cat can manage very
well thank you without fruit and veg. How?
He synthesizes Vitamin C within his own body.

A cat
is for life

By and large, our cats live longer than our dogs. The average life span of the domestic pet cat is fourteen years but, because of advances in feline nutrition and health care, it is increasing. The oldest cat on record and whose birth was fully documented, died at the age of thirty-four, but there are unverified claims of a Scottish cat reaching the ripe old age of forty-three!

David Taylor

David Taylor is a highly respected veterinary surgeon. The founder of the International Zoo Veterinary Group, he travels the world treating a wide range of animals. He is the author of many petcare books and has presented and featured in several TV shows. He is the proud owner of five Birman cats.

Acknowledgements

The publishers would like to thank the following cats for kindly allowing us to photograph them: Tupac and Tom Payne (owned by Roxy Beaujolais), Maui and Heidi (owned by Delta Rae), Angus Robertson, Pansy Puffball, Priscilla, Cool Dude, Katmandu, Scaredy Cat and Freda (owned by Natalie).
Our thanks also to the cats and staff of the Mayhew Animal Home (www.mayhewanimalhome.org, 0208 969 0178).